student study
ART NOTEBOOK

Basic
GENETICS
Second Edition

Robert F. Weaver
University of Kansas

Philip W. Hedrick
Arizona State University

Wm. C. Brown Communications, Inc.

President and Chief Executive Officer *G. Franklin Lewis*
Corporate Senior Vice President, President of WCB Manufacturing *Roger Meyer*
Corporate Senior Vice President and Chief Financial Officer *Robert Chesterman*

The credits section for this book begins on page 63 and is considered an extension of the copyright page.

Copyright © 1995 by Wm. C. Brown Communications, Inc.
All rights reserved

A Times Mirror Company

ISBN 0-697-24302-8

No part of this publication may be reproduced, stored in a retrieval system, or transmitted, in any form or by any means, electronic, mechanical, photocopying, recording, or otherwise, without the prior written permission of the publisher.

Printed in the United States of America by Wm. C. Brown Communications, Inc., 2460 Kerper Boulevard, Dubuque, IA 52001

10 9 8 7 6 5 4 3 2

TO INSTRUCTORS AND STUDENTS

This Student Study Art Notebook is free with a new textbook to all students and can be used to take notes during lectures. On each notebook page, there are two figures (sometimes one, sometimes three) faithfully reproduced from the original textbook figure. Each figure also corresponds to each of the 100 acetates available to instructors with adoption of the text.

The intention is to place a copy of the transparency acetate art in front of students (via the notebook) as the instructor uses the overhead during lectures. The advantage to the student is that he/she will be able to see all labels clearly, and take meaningful notes without having to make hurried sketches of the acetate figure.

The pages of the Art Notebook are perforated and three-hole punched, so they can be removed and placed in a personal binder for specific study and review, or to create space for additional notes.

DIRECTORY OF NOTEBOOK FIGURES

TO ACCOMPANY

WEAVER/HEDRICK, BASIC GENETICS, 2/E

Chapter 2

The Genetic Basis of the Principle of Segregation in F_2 Progeny Using Round and Wrinkled Phenotypes Figure 2.5 — 1

The Genetic Basis of the Principle of Segregation in a Testcross Figure 2.6 — 1

The Genetic Basis of the Principle of Independent Assortment in F_2 Progeny Figure 2.7 — 2

A Cross between Two Cats Doubly Heterozygous at the Long-Hair and White-Spotted Genes Figure 2.11 — 2

The Forked-Line Approach for Progeny of a Dihybrid Cross in the Garden Pea Figure 2.12 — 3

The Eight Progeny Categories and Their Probabilities for the Cross between $+vg\ Cy+\ sese$ and $+vg++\ +se$ Figure 2.14b — 3

Symbols Used in Human Pedigrees Figure 2.15 — 4

A Human Pedigree of a Recessive Trait, Showing the Inferred Genotypes Figure 2.16 — 4

A Human Pedigree for a Dominant Trait, Showing the Inferred Genotypes Figure 2.19 — 4

Chapter 3

A Cross between Red and White Four-O'clocks Showing Incomplete Dominance Figure 3.1 — 5

A Cross between Two Yellow Mice, Yielding a 2:1 Ratio of Yellow to Agouti-Colored Mice Figure 3.2 — 5

The Potential for Successful Blood Transfusions, Indicated by a + for Different ABO Donor-Recipient Combinations Figure 3.10 — 6

A Pedigree Illustrating the Pattern of Inheritance Expected for an X-Linked Recessive Gene Figure 3.15 — 6

A Pedigree Illustrating the Pattern of Inheritance Expected for an X-Linked Dominant Gene Figure 3.17 — 6

A Cross between Two Types of White Sweet Peas, Showing the 9:7 Ratio That Results in the F_2 Generation Figure 3.19 — 7

A Pedigree Illustrating a Recessive Disease Caused by Two Genes Figure 3.21 — 7

A Cross of Red- and White-Kerneled Wheat Differing at Two Genes, Showing the Five Types of F_2 Progeny Figure 3.29 — 8

The Change in Phenotypic Mean for Populations with Different Heritabilities Figure 3.34 — 9

The HLA Haplotypes and Genotypes of a Mother, Child, and Two Putative Fathers Figure 3.37 — 9

Chapter 4

The Three Major Types of Chromosomes as They Appear in Human Karyotypes Figure 4.3 — 9

The Behavior of Two Pairs of Chromosomes in Mitosis, Where Pair One Is Acrocentric (Red) and Pair Two Is Metacentric (Blue) Figure 4.7 — 10

A Schematic Representation of Meiosis for Two Chromosomes Figure 4.9 — 11

The Stages in the Production of Mature Male and Female Gametes Figure 4.11 — 12

The Chromosomal Basis of Mendel's Principle of Segregation Operating in an Rr Heterozygote Figure 4.13 — 12

The Chromosomal Basis of Mendel's Principle of Independent Assortment in an $RrYy$ Double Heterozygote Figure 4.14 — 13

Results of a Cross between a Red-Eyed Female and a White-Eyed Male Figure 4.15 — 14

Loop Formed in Chromosomal Heterozygotes for (a) Paracentric Inversion and (b) Pericentric Inversion Figure 4.24 — 14

The Most Common Types of Events Leading to Two Types of Translocations Figure 4.26 — 15

A Heterozygote for a Reciprocal Translocation and the Outcomes of Meiotic Division Figure 4.27 — 15

Comparison of Late-Prophase Banding Sequences of the Five Largest Chromosomes in Humans, Chimpanzees, Gorillas, and Orangutans Figure 4.40 — 16

Chapter 5

A Schematic of the Gametes and Progeny from a Testcross, with Coupling in the Parents Figure 5.3 — 17

The Three Possible Orders of Genes *y, w,* and *m* Produced from Double Recombination Figure 5.6 — 18
Fusion of Human and Mouse Cells to Identify the Chromosomal Location of the Gene for Thymidine Kinase Figure 5.12 — 18
The Different Consequences of Two-Stranded and Four-Stranded Crossing Over Figure 5.13 — 19
The Gametes Produced from an Attached X Heterozygous for *Bar* With No Crossing Over and from Two- and Four-Stranded Crossing Over Figure 5.16 — 20
The Ascospore Order That Results from the Three Different Types of Double Crossovers Figure 5.19 — 21
The Meiotic Products from a Single Crossover within (a) a Paracentric and (b) Pericentric Inversion Loop Figure 5.20 — 22
The Results of Normal Crossing Over and of Mispairing and Unequal Crossing Over That Produce Both Wild-Type and *Ultrabar* Chromosomes Figure 5.21 — 23

Chapter 6
The Hershey-Chase Experiment Figure 6.5 — 24
The Bases of DNA and RNA Figure 6.6 — 25
Two Examples of Nucleosides Figure 6.8 — 25
A Trinucleotide Figure 6.11 — 25
Two Models of DNA Structure Figure 6.14(b & c) — 26

Chapter 7
Two Hypotheses for DNA Replication Figure 7.2 — 27
Continuous, Semidiscontinuous, and Discontinuous Models of DNA Replication Figure 7.3 — 27
Removing Primers and Joining Nascent DNA Fragments Figure 7.6 — 28
Summary of the Mechanism of DNA Replication Figure 7.7 — 28
Two Replication Hypotheses (a) the Semiconservative Model and (b) the Conservative Model Figure 7.10 — 28
The Theta Mode of DNA Replication in *Escherichia coli* Figure 7.12 — 29
Rolling Circle Model for Phage λ DNA Replication Figure 7.21 — 30
Coping with the Gaps Left by Primer Removal Figure 7.22 — 30
Examples of Recombination. The X's Represent Crossovers between the Two Chromosomes or Parts of the Same Chromosome Figure 7.24 — 31
Mechanism of Recombination Figure 7.25 — 32

Chapter 8
RNA Polymerase-Promoter Binding Figure 8.4 — 33
Negative Control of the *lac* Operon Figure 8.7 — 33
Effects of Regulatory Mutations in the *lac* Operon in Merodiploids Figure 8.9 — 34
Temporal Control of Transcription in Phage SPO1-Infected *B. subtilis* Figure 8.12 — 35
Temporal Control of Transcription in Phage T7-Infected *E. coli* Figure 8.13 — 35
Lytic Versus Lysogenic Infection by Phage λ Figure 8.15 — 36
Temporal Control of Transcription During Lytic Infection by Phage λ Figure 8.17 — 37
The Helix-Turn-Helix Motif as a DNA-Binding Element Figure 8.20 — 37

Chapter 9
The Solenoid Model of Chromatin Folding Figure 9.6 — 38
Enhancers are Orientation- and Position-Independent Figure 9.18 — 38
Outline of Splicing. The Introns in a Gene are Transcribed along with the Exons in the Primary Transcript Figure 9.26 — 38
Polyadenylation Figure 9.31 — 39
Levels of Gene Expression in Eukaryotes Figure 9.34 — 39

Chapter 10
Amino Acid Structure Figure 10.1 — 40
Experimental Test of the Messenger Hypothesis Figure 10.9 — 41
Composition of the *E. coli* Ribosome Figure 10.11 — 42
Three-Dimensional Structure of tRNA Figure 10.18 — 42
Initiation of Translation in *E. coli* Figure 10.26 — 43
Elongation in Translation Figure 10.27 — 44
Puromycin Structure and Activity Figure 10.29 — 45
Mechanism of Suppression Figure 10.31 — 46

Chapter 11
Spontaneous Mutation Induced by Tautomerization Figure 11.10 — 47
Thymine Dimers Figure 11.20 — 48
Reversion of Frameshift Mutation Figure 11.22 — 48
The Ames Test (a) Outline of the Procedure (b) Data from a Test Figure 11.23 — 49

Chapter 12
Tracking Transposition with Antibiotic Resistance Markers Figure 12.2 — 49
Deletion and Inversion Promoted by Transposons Figure 12.6 — 50
Retrovirus Replication Cycle Figure 12.10 — 50

Chapter 13
Model for F$^+$ → Hfr Conversion Figure 13.6 — 51

The F Plasmid Portion of an Hfr Chromosome Determines the Direction and Order of Host Gene Transfer Figure 13.8 — **52**
Transfer of F-*lac* to an F⁻ *lac*⁻ Cell Figure 13.12 — **53**
Creation of λ Transducing Phages Figure 13.15 — **53**
Generalized Transduction Figure 13.16 — **53**
Three-Factor Cross to Determine the Order of Genes A, B, and C Figure 13.18 — **54**

Chapter 14
The Race between Transcription Factors and Histones for the 5S rRNA Control Region Figure 14.11 — **54**
Rearrangement of an Antibody Light Chain Gene Figure 14.23 — **55**

Chapter 15
Cloning Foreign DNA Using the *Pst*I Site of pBR322 Figure 15.3 — **55**
Cloning in Charon 4 Figure 15.6 — **56**
Forming a Fusion Protein in λ gt11 Figure 15.15 — **56**
Site-Directed Mutagenesis Using an Oligonucleotide Figure 15.17 — **57**
Southern Blotting Figure 15.18 — **58**
Amplifying DNA by Polymerase Chain Reaction (PCR) Figure B15.1.2. — **59**
The Sanger Dideoxy Method of DNA Sequencing Figure 15.22 — **60**

Chapter 16
Frequencies of Inversions on the Third Chromosome in *Drosophila pseudoobscura* That Have Taken Place in the Capitan Area of New Mexico Figure 16.2 — **61**
Hardy-Weinberg Proportions as Generated from the Random Union of Gametes, Using a Unit Square Figure 16.6 — **61**

Chapter 17
The Change in Allelic Frequencies Where Both Mutation to and Selection Against Recessives Occur Figure 17.1 — **62**
A Phylogeny of Humans, Chimpanzees, Gorillas, and Orangutans, Based on Their Base Sequences in the β-Globin Region Figure 17.10 — **62**

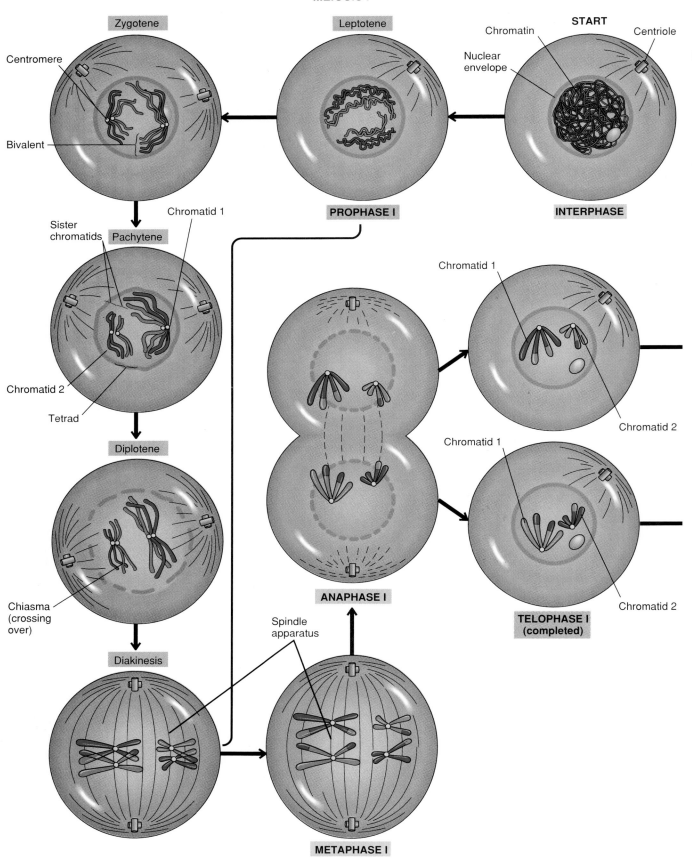

Correction of the art found on page 6, Panel 4 of *Basic Genetics*, second edition by Weaver and Hedrick.

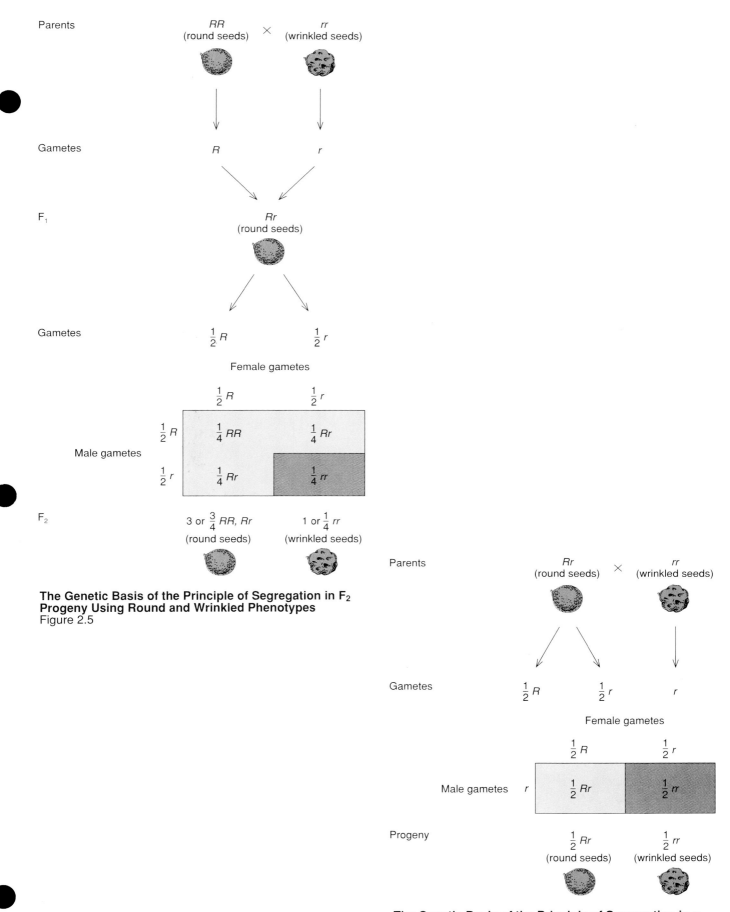

The Genetic Basis of the Principle of Segregation in F_2 Progeny Using Round and Wrinkled Phenotypes
Figure 2.5

The Genetic Basis of the Principle of Segregation in a Testcross
Figure 2.6

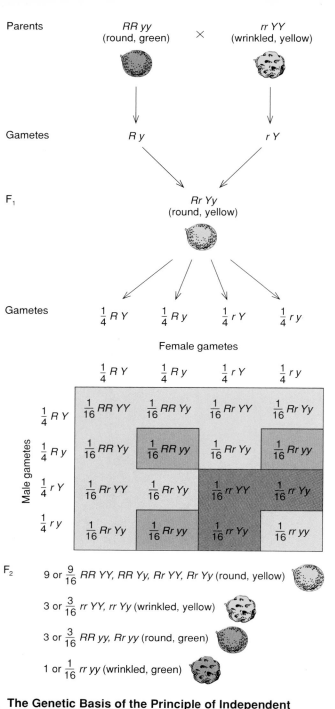

The Genetic Basis of the Principle of Independent Assortment in F_2 Progeny
Figure 2.7

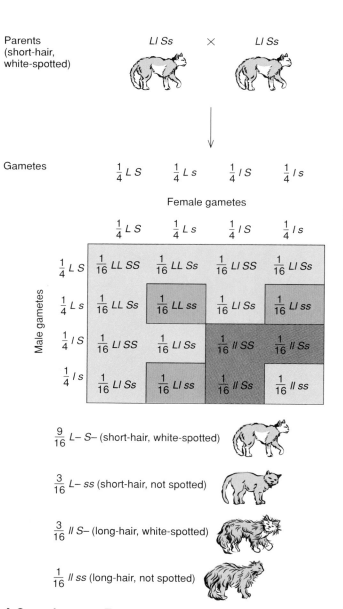

A Cross between Two Cats Doubly Heterozygous at the Long-Hair and White-Spotted Genes
Figure 2.11

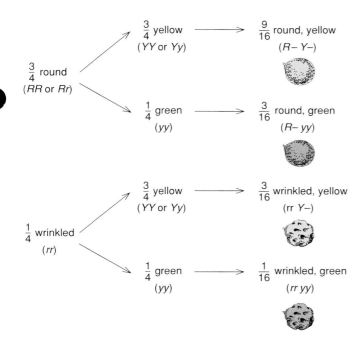

The Forked-Line Approach for Progeny of a Dihybrid Cross in the Garden Pea
Figure 2.12

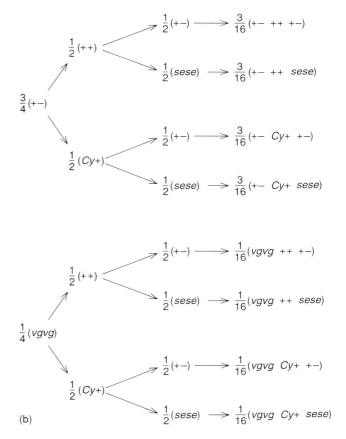

(b)

The Eight Progeny Categories and Their Probabilities for the Cross between +vg Cy+ sese and +vg++ +se
Figure 2.14b

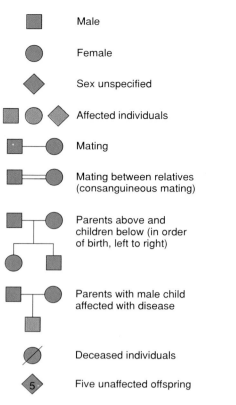

Symbols Used in Human Pedigrees
Figure 2.15

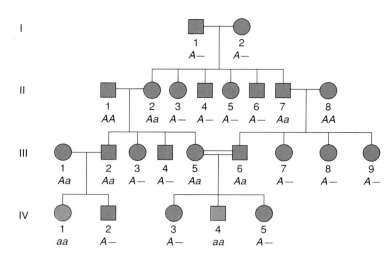

A Human Pedigree of a Recessive Trait, Showing the Inferred Genotypes
Figure 2.16

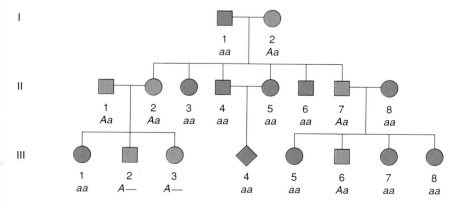

A Human Pedigree for a Dominant Trait, Showing the Inferred Genotypes
Figure 2.19

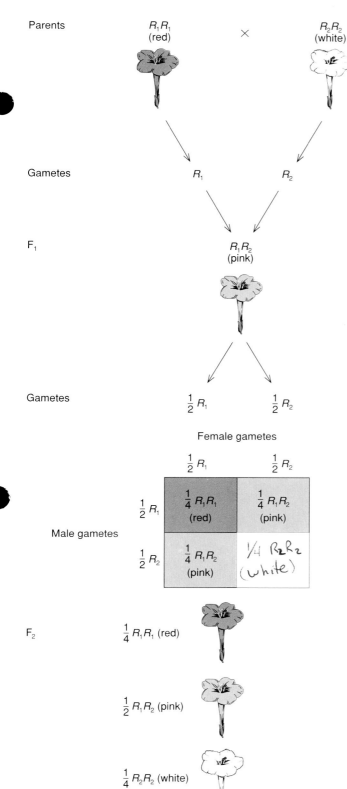

A Cross between Red and White Four-O'clocks Showing Incomplete Dominance
Figure 3.1

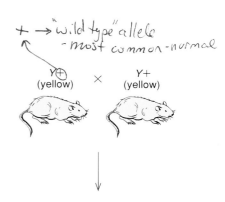

+ → "wild type" allele
- most common - normal

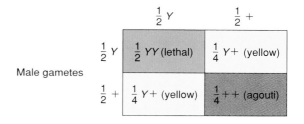

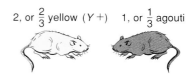

A Cross between Two Yellow Mice, Yielding a 2:1 Ratio of Yellow to Agouti-Colored Mice
Figure 3.2

Donor		Recipient			
Genotype	Phenotype	A (anti-B)	B (anti-A)	AB	O (anti-A and anti-B)
I^AI^A, I^AI^O	A	+	A, anti-A	+	A, anti-A
I^BI^B, I^BI^O	B	B, anti-B	+	+	B, anti-B
I^AI^B	AB	B, anti-B	A, anti-A	+	A, anti-A B, anti-B
I^OI^O	O	+	+	+	+

The Potential for Successful Blood Transfusions, Indicated by a + for Different ABO Donor-Recipient Combinations
Figure 3.10

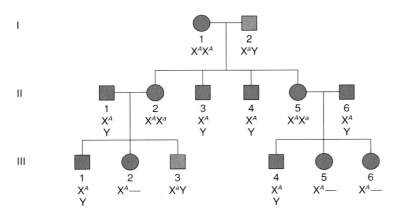

A Pedigree Illustrating the Pattern of Inheritance Expected for an X-Linked Recessive Gene
Figure 3.15

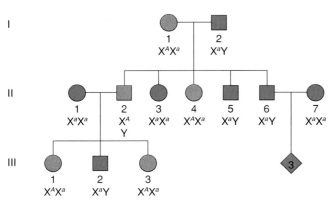

A Pedigree Illustrating the Pattern of Inheritance Expected for an X-Linked Dominant Gene
Figure 3.17

6

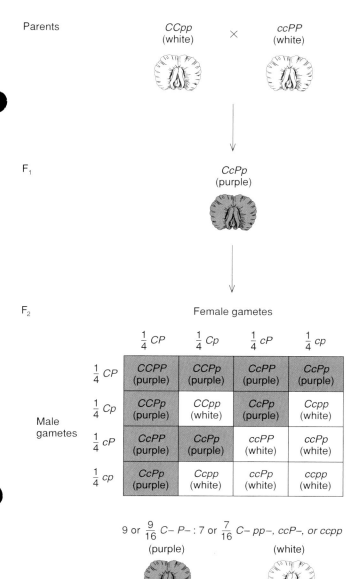

A Cross between Two Types of White Sweet Peas, Showing the 9:7 Ratio That Results in the F₂ Generation
Figure 3.19

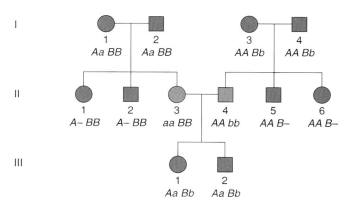

A Pedigree Illustrating a Recessive Disease Caused by Two Genes
Figure 3.21

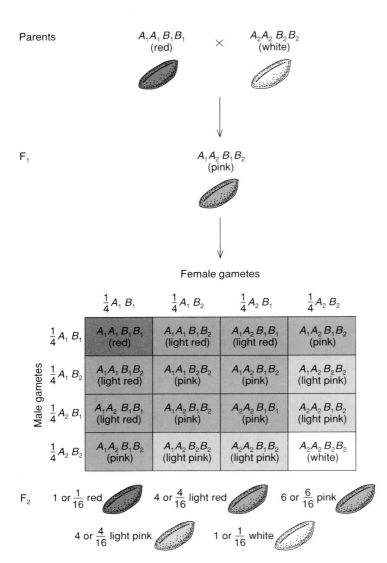

A Cross of Red- and White-Kerneled Wheat Differing at Two Genes, Showing the Five Types of F$_2$ Progeny
Figure 3.29

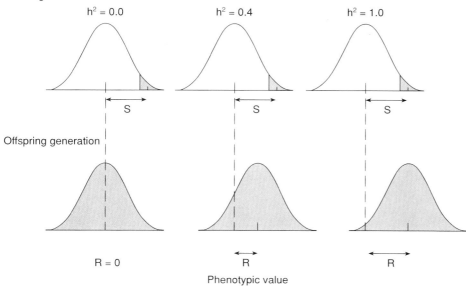

The Change in Phenotypic Mean for Populations with Different Heritabilities
Figure 3.34

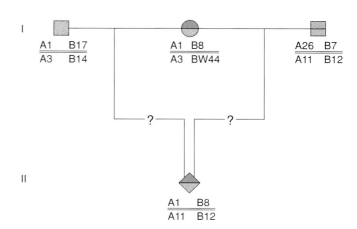

The HLA Haplotypes and Genotypes of a Mother, Child, and Two Putative Fathers
Figure 3.37

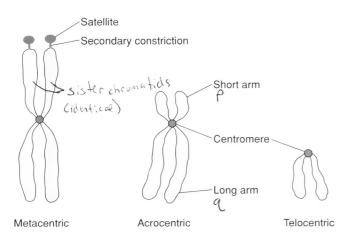

The Three Major Types of Chromosomes as They Appear in Human Karyotypes
Figure 4.3

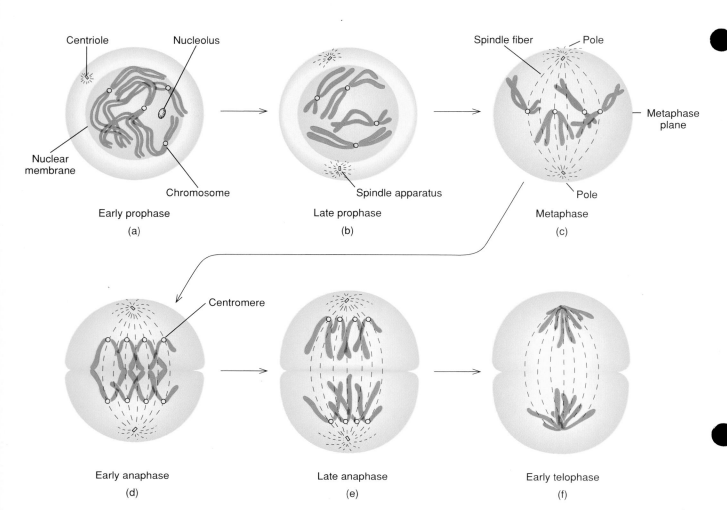

The Behavior of Two Pairs of Chromosomes in Mitosis, Where Pair One Is Acrocentric (Red) and Pair Two Is Metacentric (Blue)
Figure 4.7

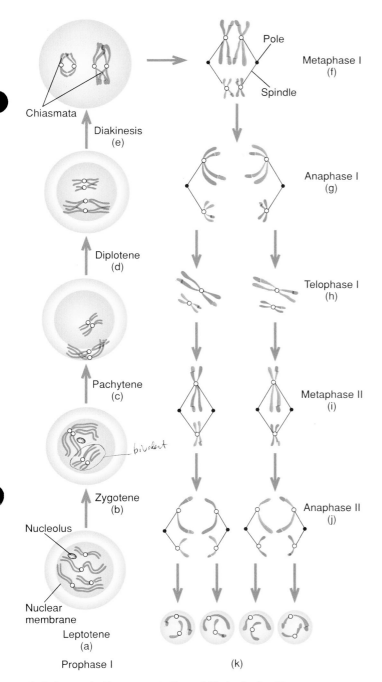

A Schematic Representation of Meiosis for Two Chromosomes
Figure 4.9

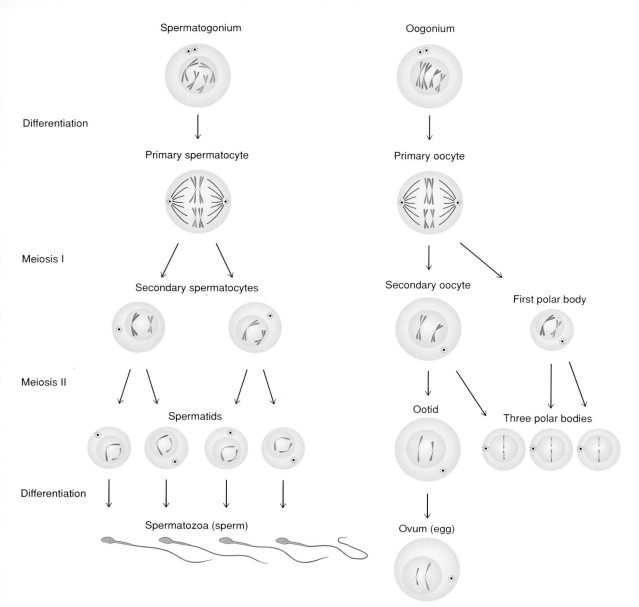

The Stages in the Production of Mature Male and Female Gametes
Figure 4.11

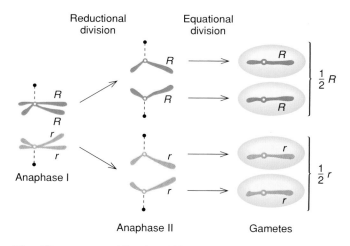

The Chromosomal Basis of Mendel's Principle of Segregation Operating in an *Rr* Heterozygote
Figure 4.13

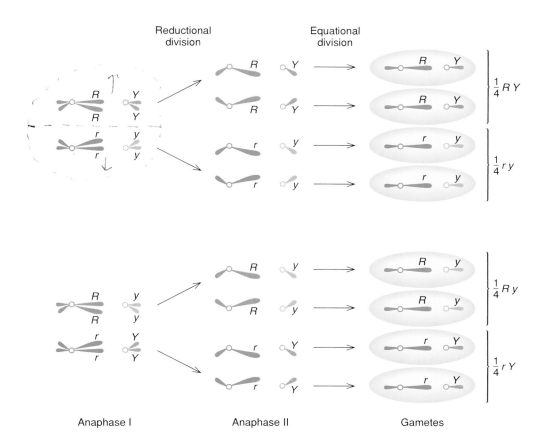

The Chromosomal Basis of Mendel's Principle of Independent Assortment in an *RrYy* Double Heterozygote
Figure 4.14

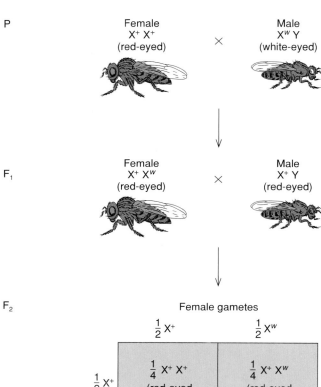

Results of a Cross between a Red-Eyed Female and a White-Eyed Male
Figure 4.15

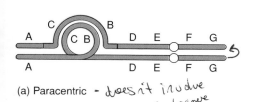
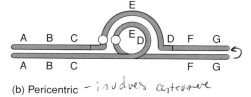

(a) Paracentric — does it involve centromere

(b) Pericentric — involves centromere

Loop Formed in Chromosomal Heterozygotes for (a) Paracentric Inversion and (b) Pericentric Inversion
Figure 4.24

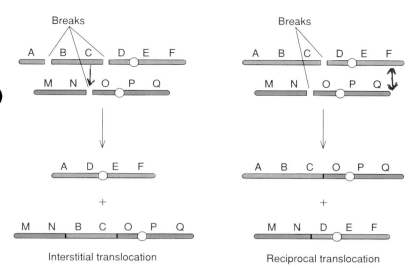

The Most Common Types of Events Leading to Two Types of Translocations
Figure 4.26

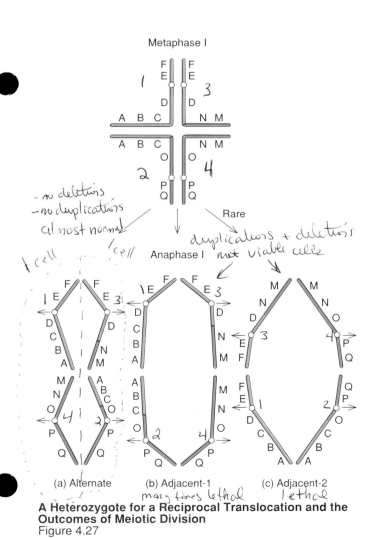

A Heterozygote for a Reciprocal Translocation and the Outcomes of Meiotic Division
Figure 4.27

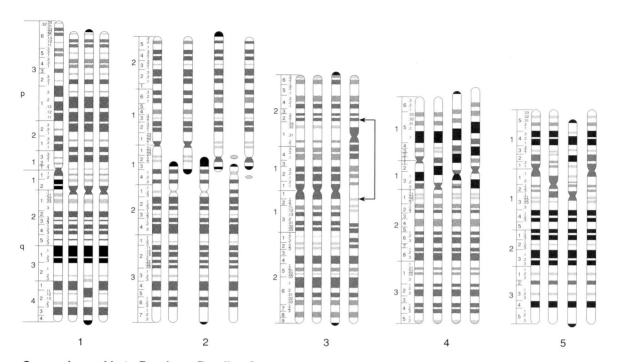

Comparison of Late-Prophase Banding Sequences of the Five Largest Chromosomes in Humans, Chimpanzees, Gorillas, and Orangutans
Figure 4.40

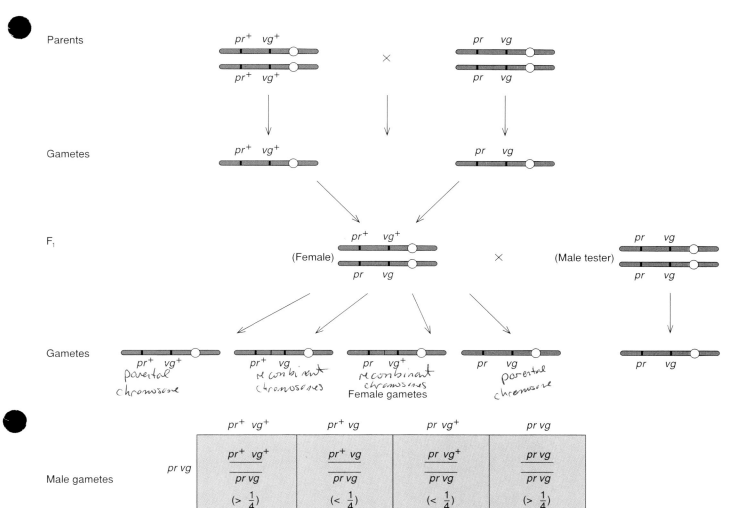

A Schematic of the Gametes and Progeny from a Testcross, with Coupling in the Parents
Figure 5.3

Drosophila Linkage –

pr^+ – normal eye color – red
pr – purple eye
vg^+ – normal wings
vg – vestigial wings – useless

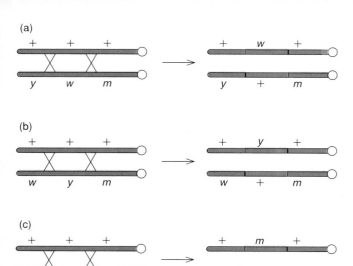

The Three Possible Orders of Genes *y*, *w*, and *m* Produced from Double Recombination
Figure 5.6

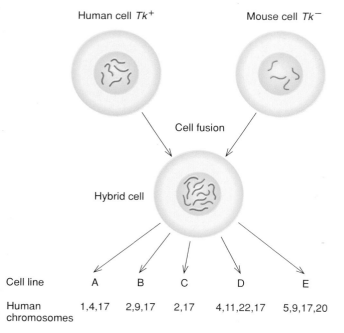

Fusion of Human and Mouse Cells to Identify the Chromosomal Location of the Gene for Thymidine Kinase
Figure 5.12

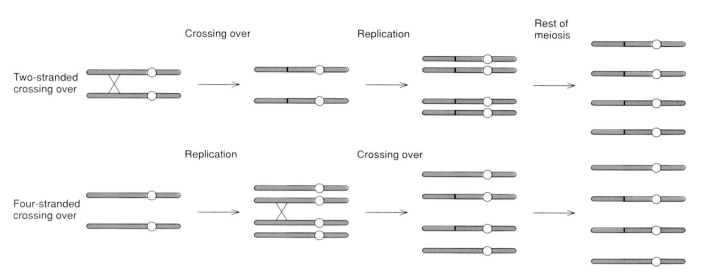

The Different Consequences of Two-Stranded and Four-Stranded Crossing Over
Figure 5.13

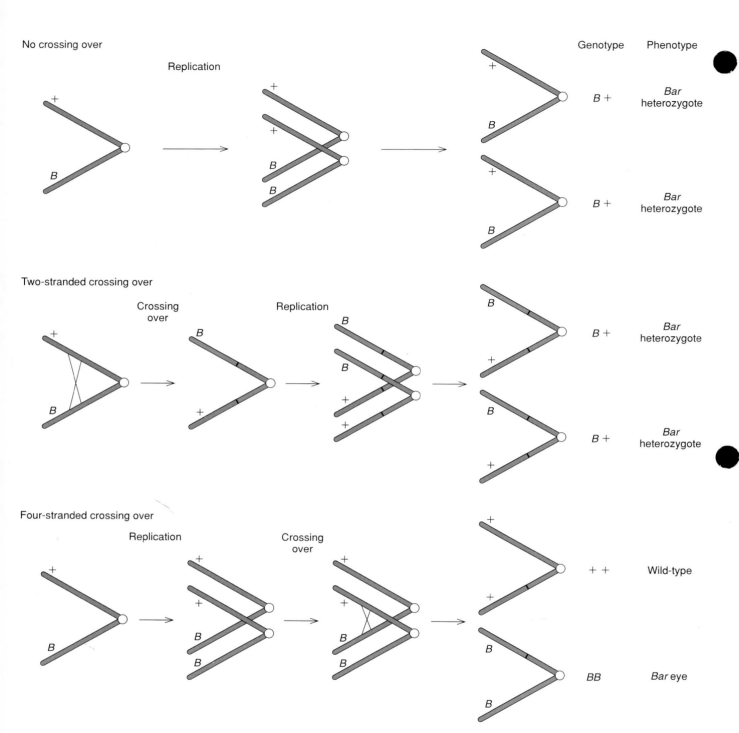

The Gametes Produced from an Attached X Heterozygous for *Bar* With No Crossing Over and from Two- and Four-Stranded Crossing Over
Figure 5.16

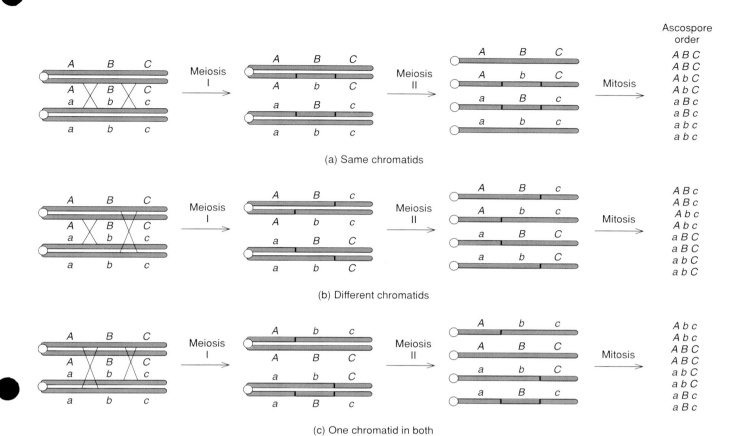

The Ascospore Order That Results from the Three Different Types of Double Crossovers
Figure 5.19

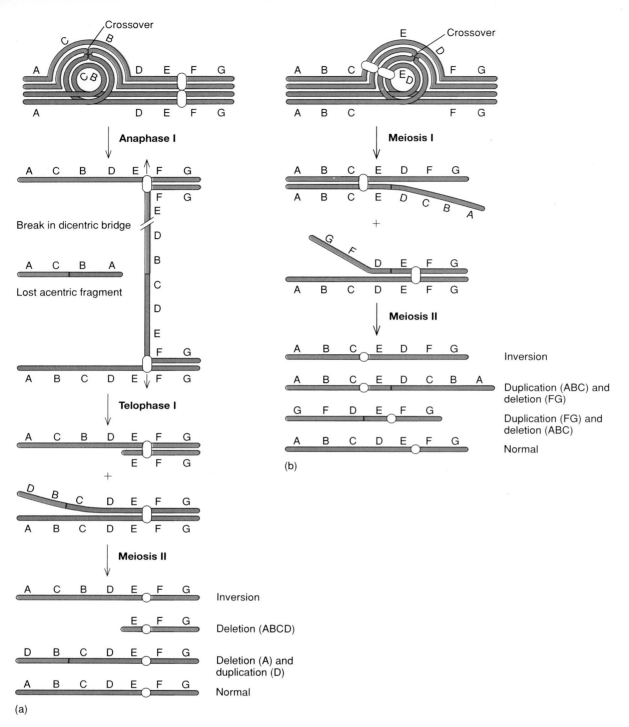

The Meiotic Products from a Single Crossover within (a) a Paracentric and (b) Pericentric Inversion Loop
Figure 5.20

Normal pairing and normal crossing over

Mispairing and unequal crossing over

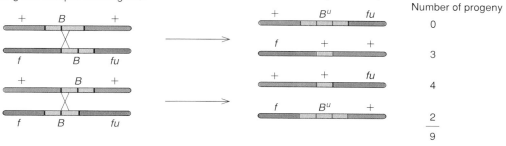

	Number of progeny
	0
	3
	4
	2/9

The Results of Normal Crossing Over and of Mispairing and Unequal Crossing Over That Produce Both Wild-Type and *Ultrabar* Chromosomes
Figure 5.21

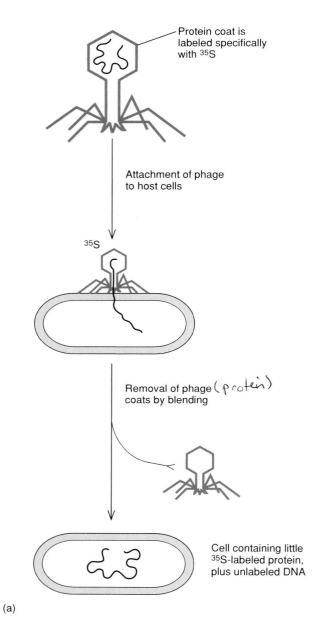
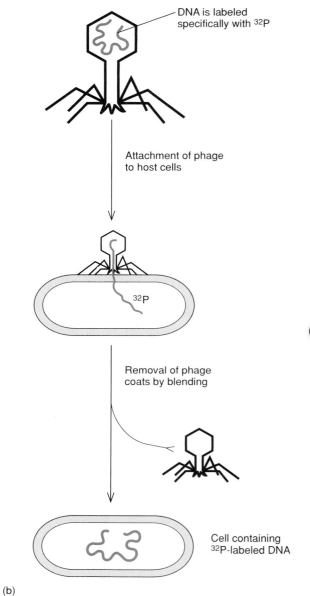

The Hershey-Chase Experiment
Figure 6.5

In search of the Fountain of Youth...

The hunt for the mythical Fountain of Youth has been going on for centuries, centuries. Yet it may be just around the corner for the next generation of scientists—that's you.

In the **Gene Game**, *you are the scientist in search of the gene that controls the aging process.*

Luckily, you have the power of interactive, easy-to-use, affordable Macintosh software.

Gene Game

by William Sofer
The State University of New Jersey—Rutgers
1995 • Macintosh software
ISBN 0-697-24893-5

from your laboratory bench.

This is your challenge
Work through lab protocols to clone a fictitious "fountain of youth" gene. These are the tools you'll use in your investigation: your own critical-thinking skills, the scientific method, and cutting-edge software.

- For each step—and misstep—in the process, you'll receive direct feedback and hints in the "lab manual" contained within the program.

- The steps and protocol taken in the lab are automatically recorded in the program's "lab notebook" that you can refer to or print out.

System requirements.
- *Macintosh computer*
- *2 MB of RAM*
- *1 MB free hard disk space*
- *System 6.07 or newer*
- *Color monitor with 13-inch or larger screen*
- *Printer (optional)*

Order now.
Contact your bookstore, or call Wm. C. Brown Publishers: 800-338-5578.

Wm. C. Brown Publishers
A Division of Wm. C. Brown Communications, Inc.

LET US HELP YOU WITH YOUR PROBLEMS.

Genetics Problem Solving Guide, Second Edition
by William R. Wellnitz, *Augusta College*
1995 • 160 pages (approx.) • paperback
ISBN 0-697-13739-2

Let this guide walk you through the logical steps involved in solving genetics problems. It reflects the latest advances in a dynamic field and incorporates helpful features like these:

- *What you need to know.* Essential Concepts, lettered and highlighted for easy identification, include *worked-out Examples* and additional information.
- *What else you need to know.* Notes in many chapters elaborate on important concepts.
- *Keep this in mind.* Chapter Summary alerts you to the key ideas to keep in mind when working through the problems pertinent to that chapter's lettered concepts.
- *Practice, practice, practice.* End-of-chapter *Practice Problems* incorporate all chapter concepts and vary in difficulty.
- *Want a challenge?* Try *Master Problems*, which often involve concepts from more than one chapter.
- *Here's the answer.* Answers to the Practice Problems describe the steps and concepts involved in solving the problem.

Compendium of Problems in Genetics
by John Kuspira and Ramesh Bhambhani,
both of the University of Alberta
1994 • 368 pages • plastic comb
ISBN 0-697-16734-8

Whether you're working at a basic or an advanced level, you'll find these problems current, logical, and thought-provoking.

- *It's relevant.* Approximately half of the problems are based on *actual experimental data from "classic" papers* published in the field.
- *It's rational.* End-of-book *solutions*, provided to more than one-third of the problems, reveal the *reasoning behind the analytical steps* used in reaching solutions.
- *It's illustrative.* *Realistic, informative diagrams* illuminate the material investigated.

Student Study Guide in Genetics, Second Edition
by Ken Zwicker, *Whitehead Institute*
1993 • 128 pages • wire coil
ISBN 0-697-13725-2

Each chapter in this clearly illustrated guide opens with *Key Concepts and Terms*, covers a range of *Study Questions*, and ends with *Answers*. Material reflects current thought, techniques, and technology.

PROBLEMS SOLVED.
Contact your bookstore now to place your order. Or call Wm. C. Brown Publishers' Customer Service Department: **800-228-0459.**

WCB Wm. C. Brown Publishers
A Division of Wm. C. Brown Communications, Inc.

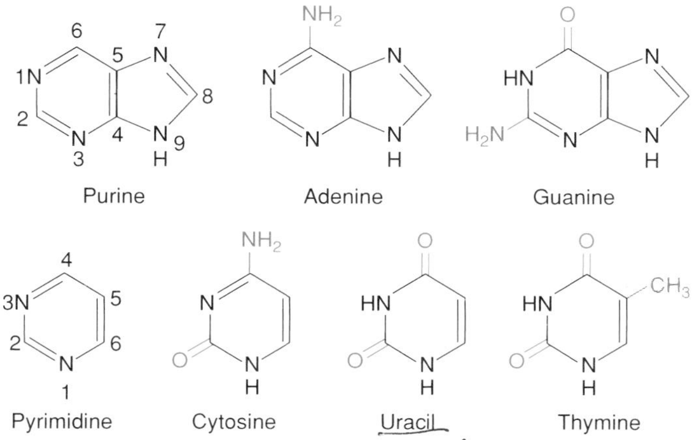

Purine
Adenine
Guanine
Pyrimidine
Cytosine
Uracil
Thymine

RNA only

The Bases of DNA and RNA
Figure 6.6

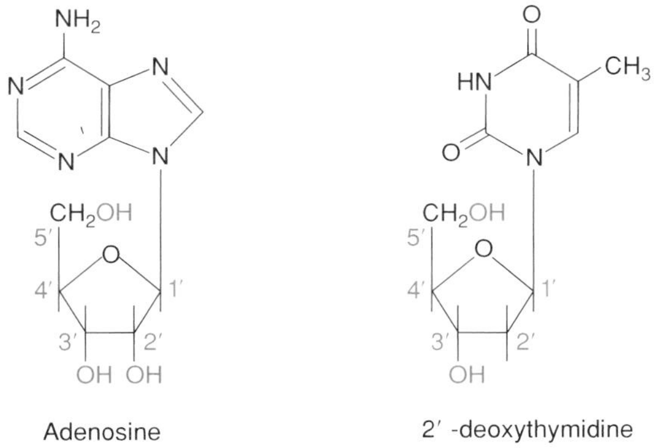

Adenosine
2′ -deoxythymidine

Two Examples of Nucleosides
Figure 6.8

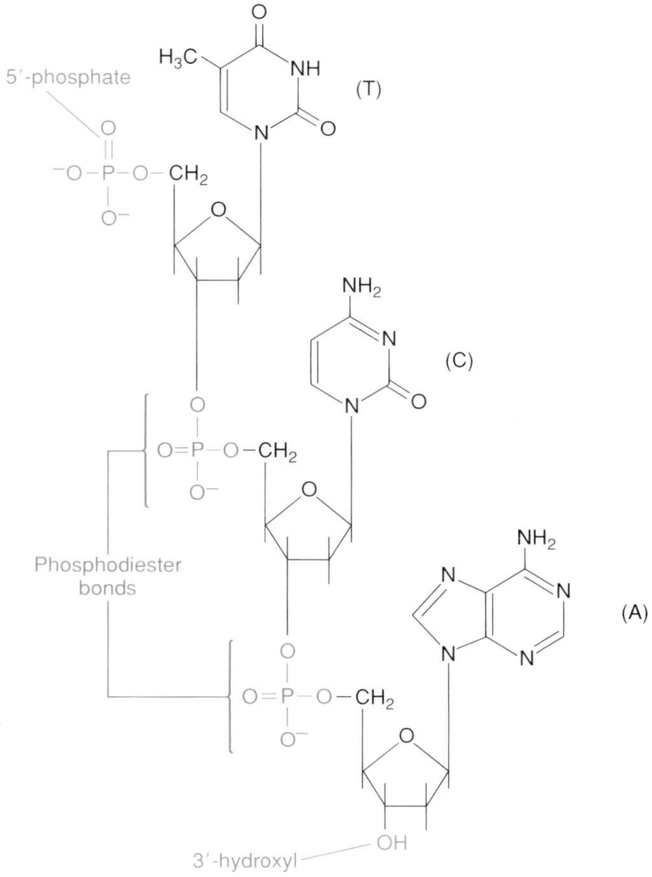

5′-phosphate
(T)
(C)
(A)
Phosphodiester bonds
3′-hydroxyl

A Trinucleotide
Figure 6.11

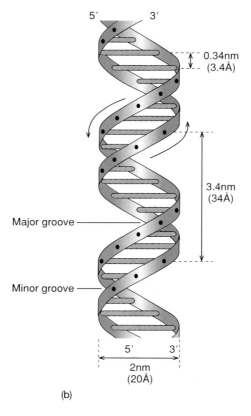

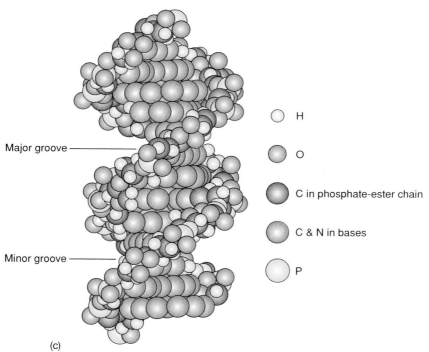

Two Models of DNA Structure
Figure 6.14(b & c)

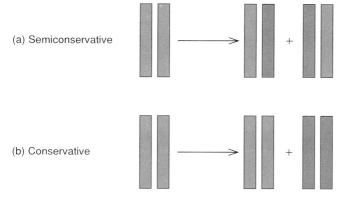

Two Hypotheses for DNA Replication
Figure 7.2

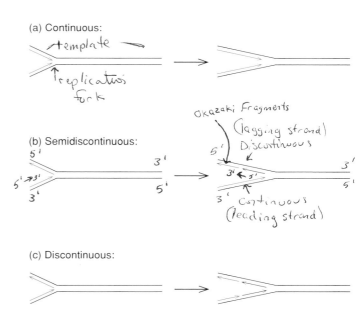

Continuous, Semidiscontinuous, and Discontinuous Models of DNA Replication
Figure 7.3

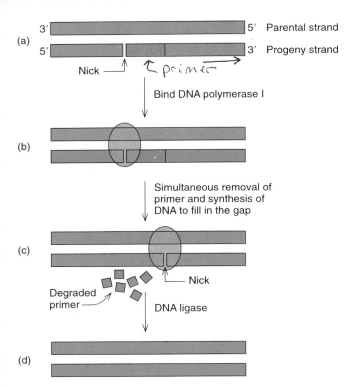

Removing Primers and Joining Nascent DNA Fragments
Figure 7.6

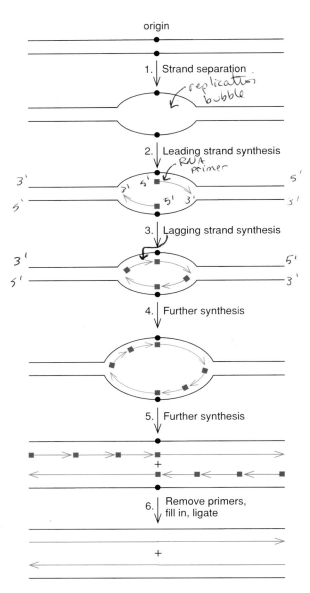

Summary of the Mechanism of DNA Replication
Figure 7.7

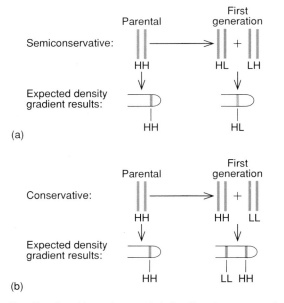

Two Replication Hypotheses (a) the Semiconservative Model and (b) the Conservative Model
Figure 7.10

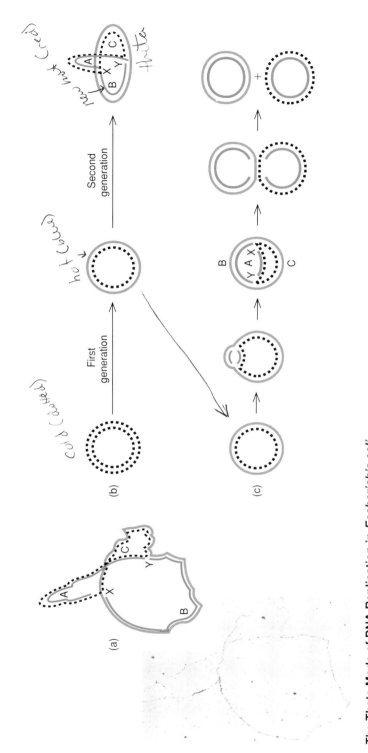

The Theta Mode of DNA Replication in *Escherichia coli*
Figure 7.12

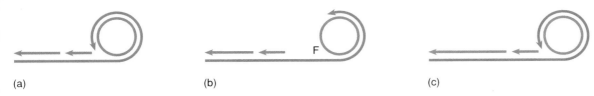

Rolling Circle Model for Phage λ DNA Replication
Figure 7.21

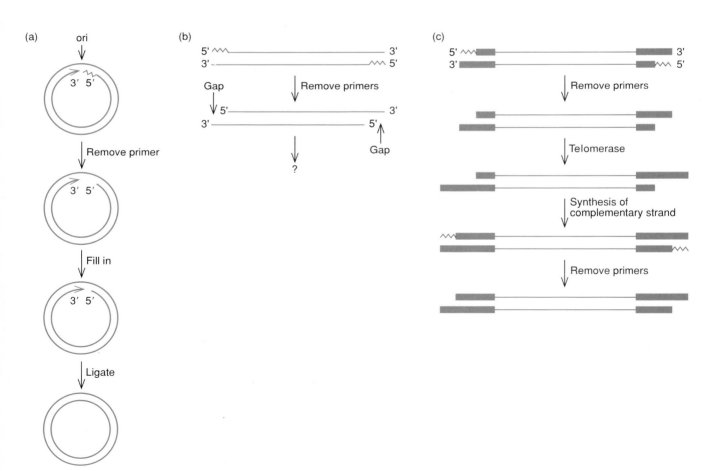

Coping with the Gaps Left by Primer Removal
Figure 7.22

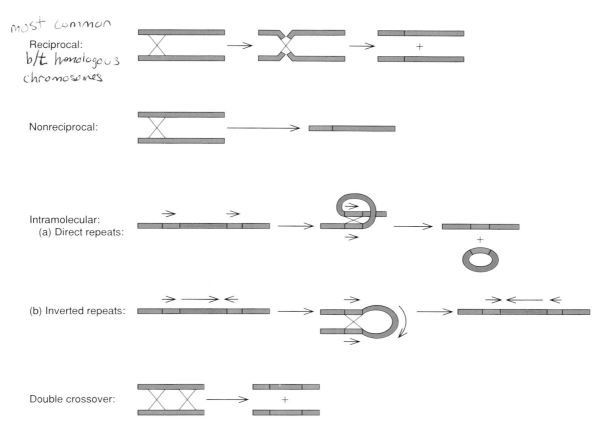

Examples of Recombination. The X's Represent Crossovers between the Two Chromosomes or Parts of the Same Chromosome
Figure 7.24

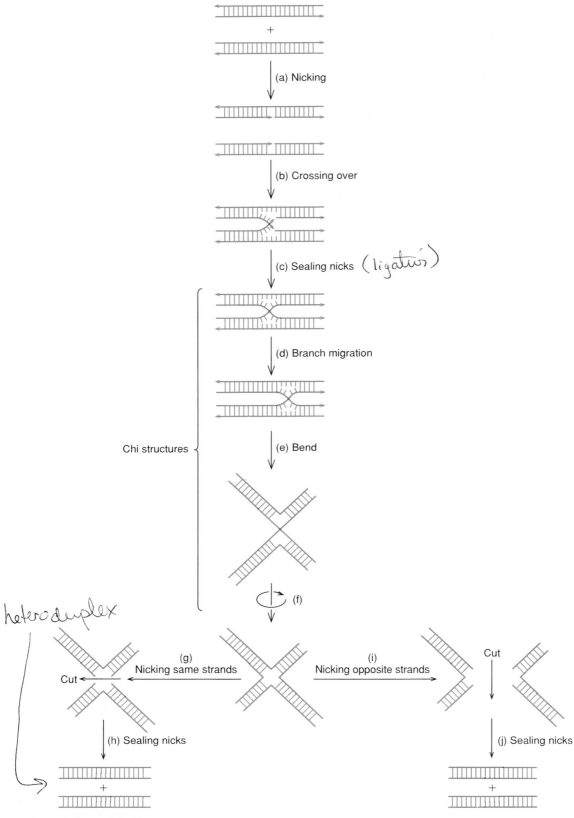

Mechanism of Recombination
Figure 7.25

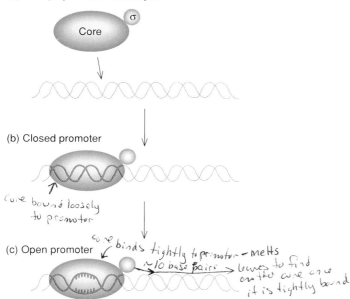

RNA Polymerase-Promoter Binding
Figure 8.4

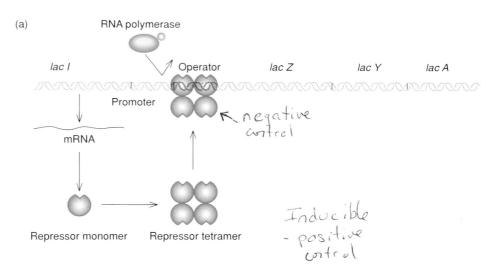

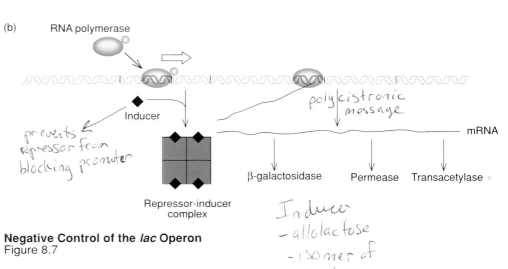

Negative Control of the *lac* Operon
Figure 8.7

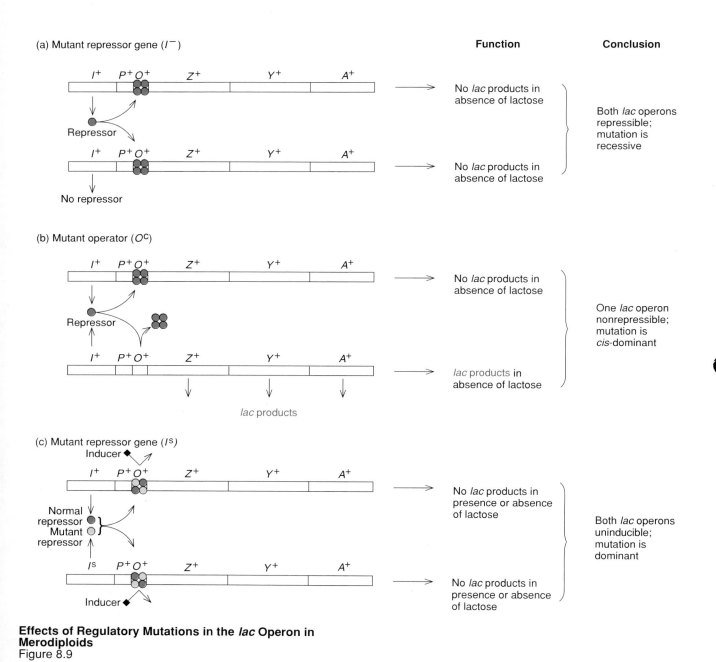

Effects of Regulatory Mutations in the *lac* Operon in Merodiploids
Figure 8.9

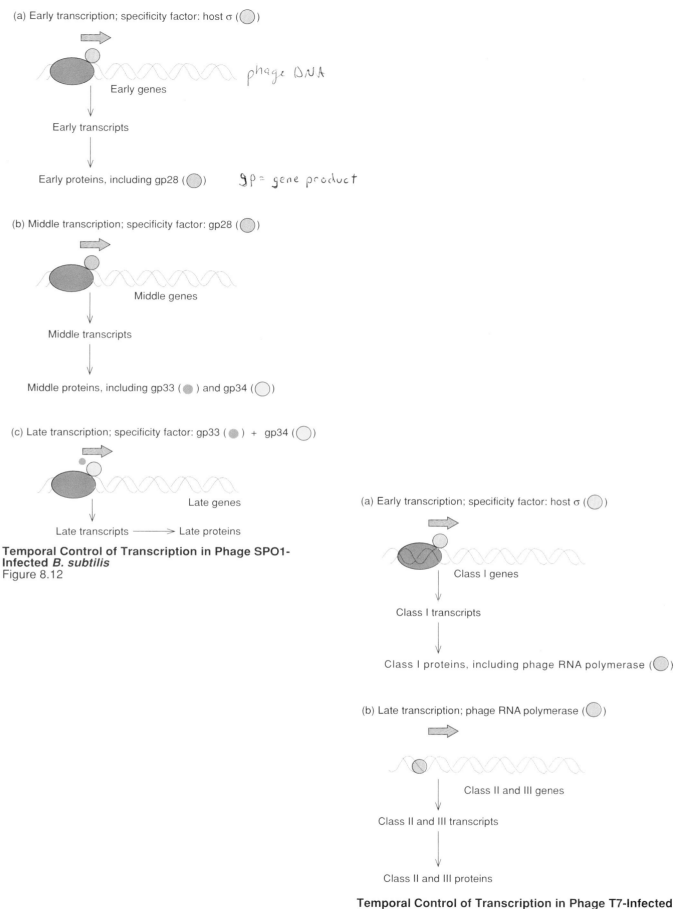

(a) Early transcription; specificity factor: host σ (○)

phage DNA

Early genes

Early transcripts

Early proteins, including gp28 (○)   gp = gene product

(b) Middle transcription; specificity factor: gp28 (○)

Middle genes

Middle transcripts

Middle proteins, including gp33 (●) and gp34 (○)

(c) Late transcription; specificity factor: gp33 (●) + gp34 (○)

Late genes

Late transcripts ⟶ Late proteins

Temporal Control of Transcription in Phage SPO1-Infected *B. subtilis*
Figure 8.12

(a) Early transcription; specificity factor: host σ (○)

Class I genes

Class I transcripts

Class I proteins, including phage RNA polymerase (○)

(b) Late transcription; phage RNA polymerase (○)

Class II and III genes

Class II and III transcripts

Class II and III proteins

Temporal Control of Transcription in Phage T7-Infected *E. coli*
Figure 8.13

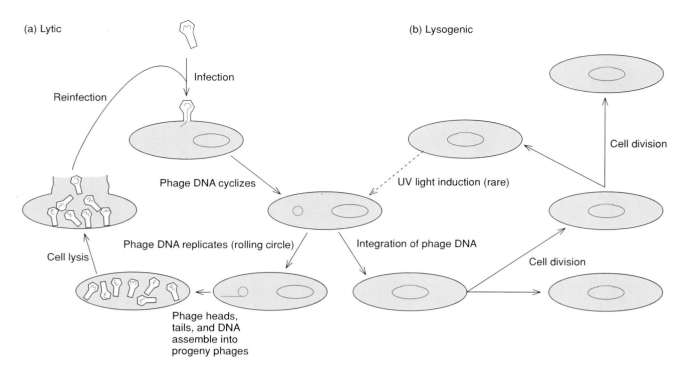

Lytic Versus Lysogenic Infection by Phage λ
Figure 8.15

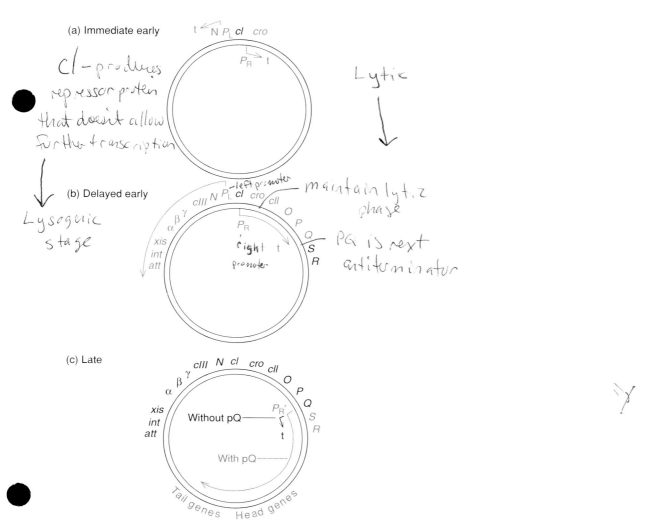

cI - produces repressor protein that doesn't allow further transcription

↓

Lysogenic stage

Lytic
↓

—left promoter — maintain lyt. z phase

PQ is next antiterminator

Temporal Control of Transcription During Lytic Infection by Phage λ
Figure 8.17

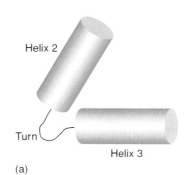

The Helix-Turn-Helix Motif as a DNA-Binding Element
Figure 8.20

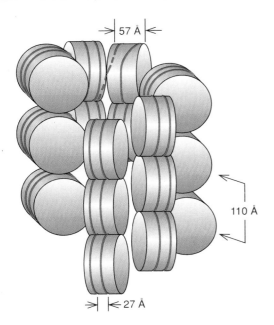

The Solenoid Model of Chromatin Folding
Figure 9.6

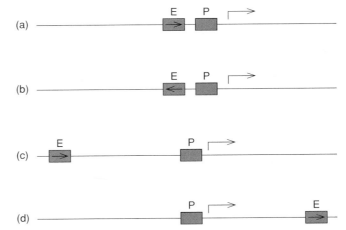

Enhancers are Orientation- and Position-Independent
Figure 9.18

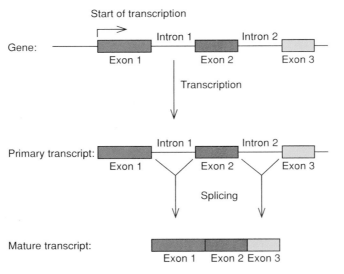

Outline of Splicing. The Introns in a Gene are Transcribed along with the Exons in the Primary Transcript
Figure 9.26

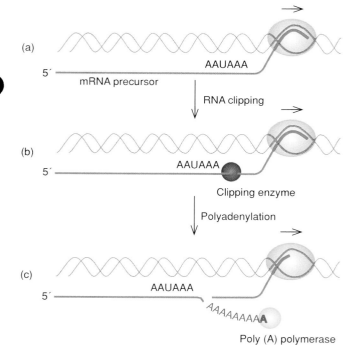

Polyadenylation
Figure 9.31

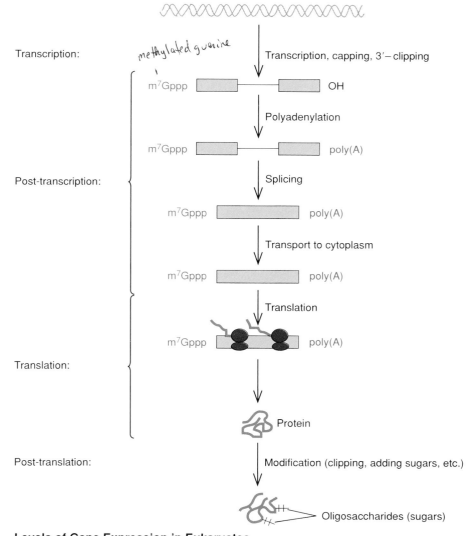

Levels of Gene Expression in Eukaryotes
Figure 9.34

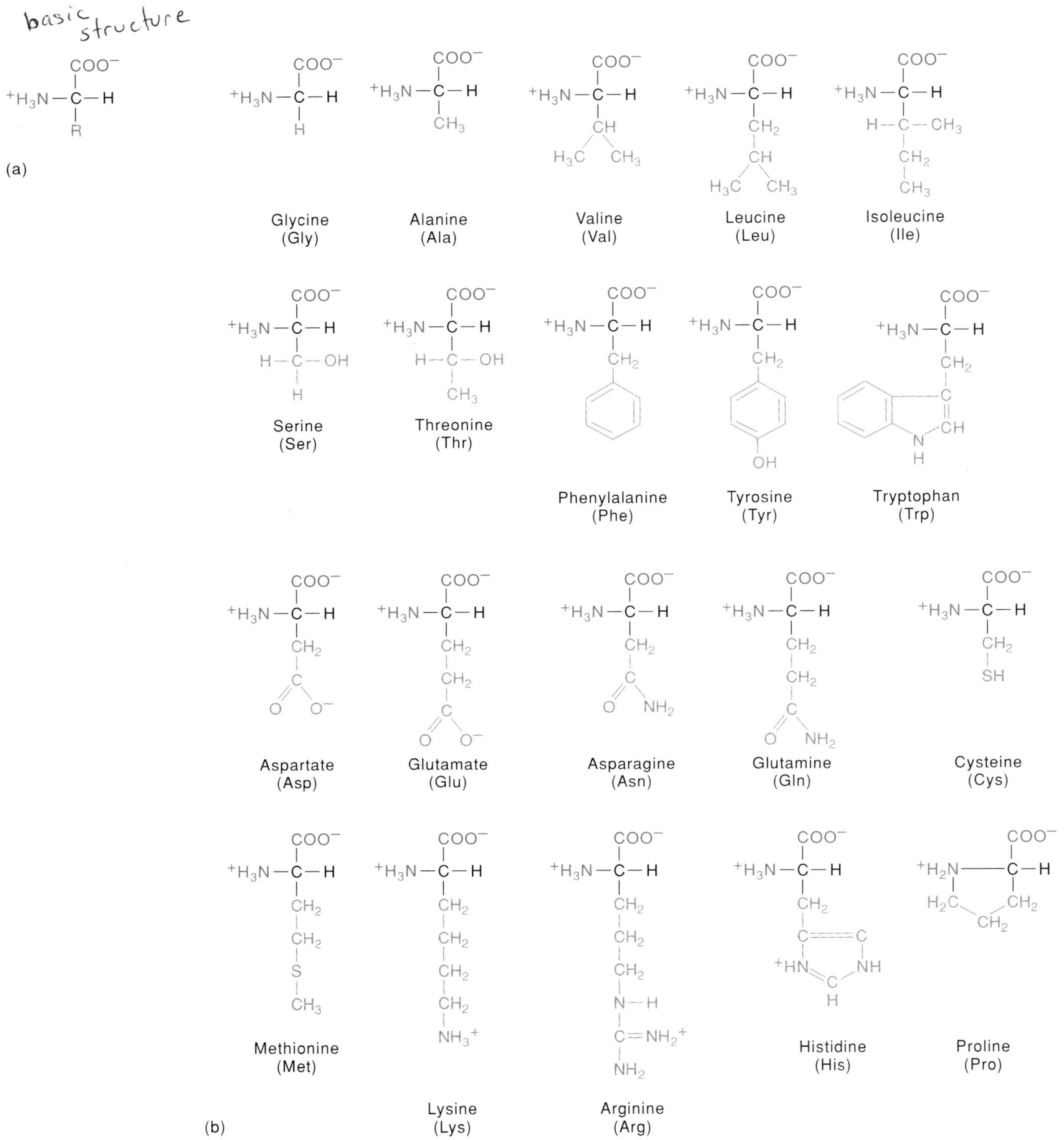

basic structure

(a)

Glycine (Gly) · Alanine (Ala) · Valine (Val) · Leucine (Leu) · Isoleucine (Ile)

Serine (Ser) · Threonine (Thr) · Phenylalanine (Phe) · Tyrosine (Tyr) · Tryptophan (Trp)

Aspartate (Asp) · Glutamate (Glu) · Asparagine (Asn) · Glutamine (Gln) · Cysteine (Cys)

Methionine (Met) · Lysine (Lys) · Arginine (Arg) · Histidine (His) · Proline (Pro)

(b)

Amino Acid Structure
Figure 10.1

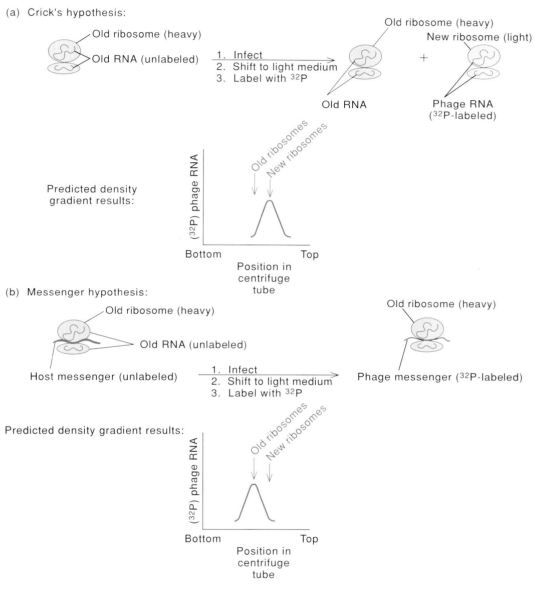

Experimental Test of the Messenger Hypothesis
Figure 10.9

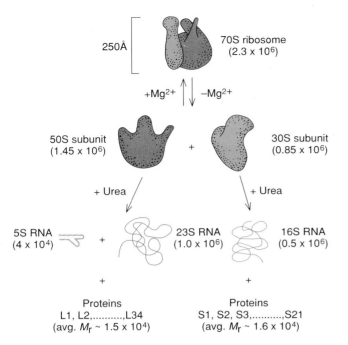

Composition of the *E. coli* Ribosome
Figure 10.11

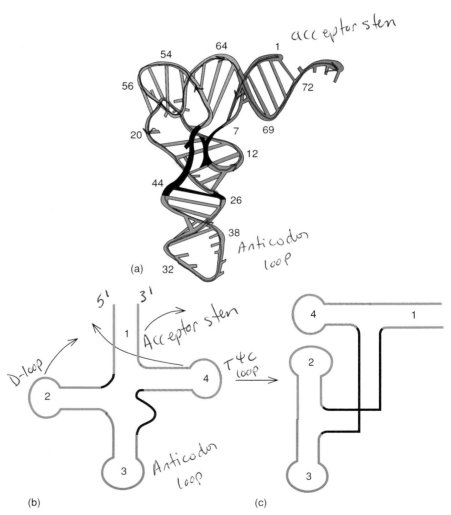

Three-Dimensional Structure of tRNA
Figure 10.18

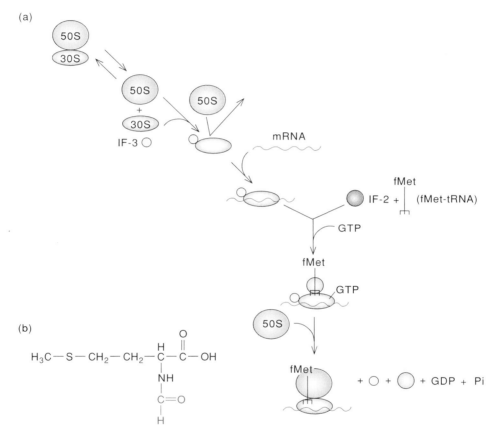

Initiation of Translation in *E. coli*
Figure 10.26

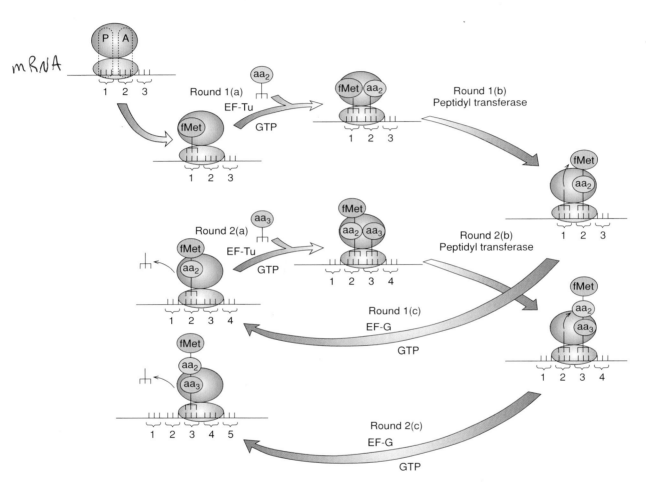

Elongation in Translation
Figure 10.27

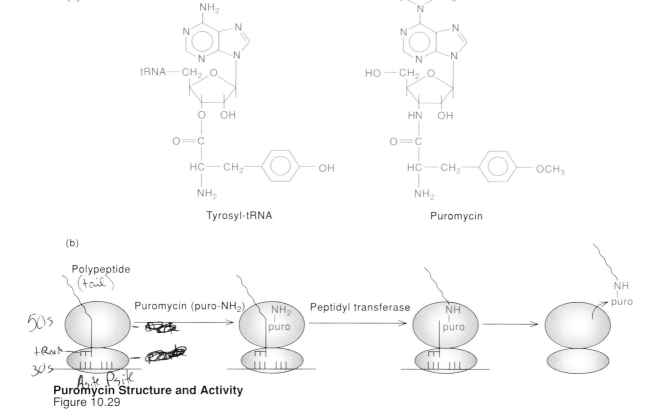

Puromycin Structure and Activity
Figure 10.29

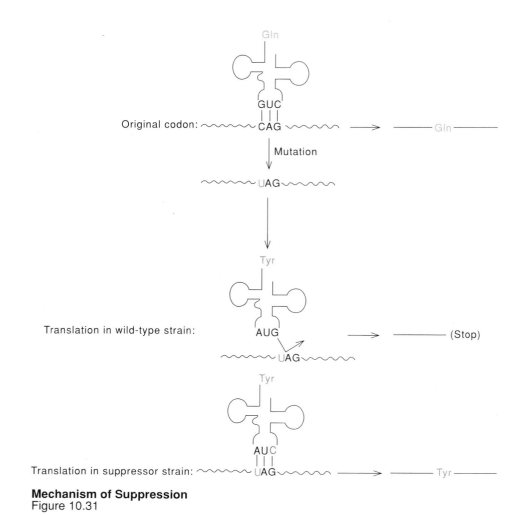

Mechanism of Suppression
Figure 10.31

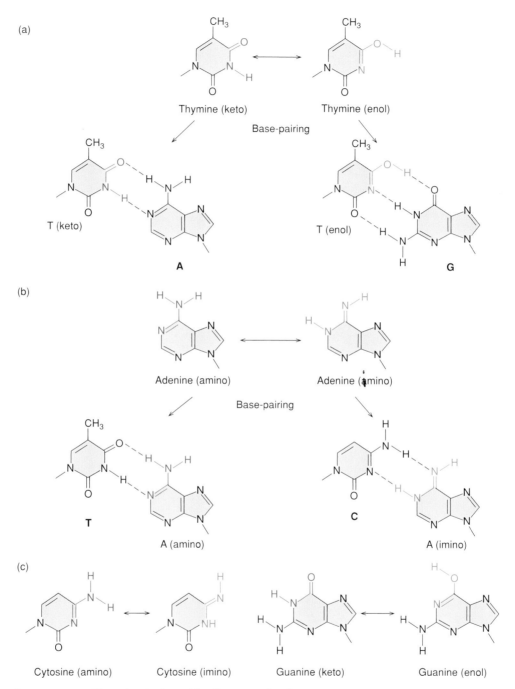

Spontaneous Mutation Induced by Tautomerization
Figure 11.10

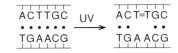

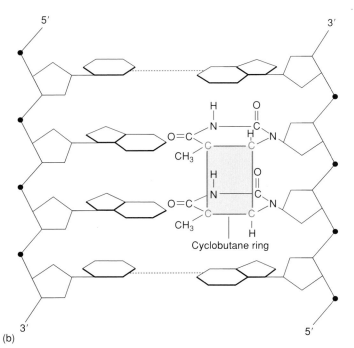

Thymine Dimers
Figure 11.20

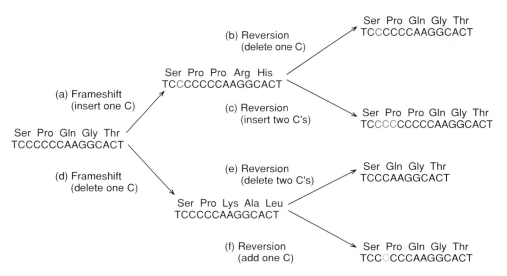

Reversion of Frameshift Mutation
Figure 11.22

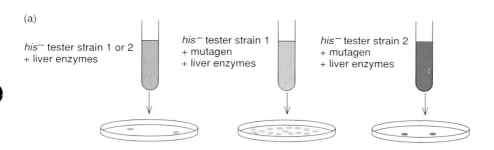

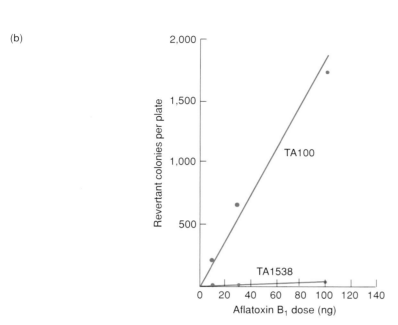

The Ames Test (a) Outline of the Procedure (b) Data from a Test
Figure 11.23

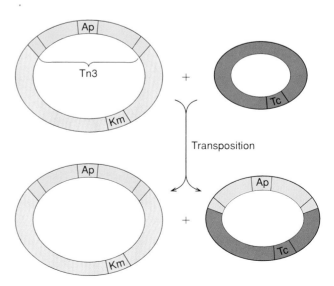

Tracking Transposition with Antibiotic Resistance Markers
Figure 12.2

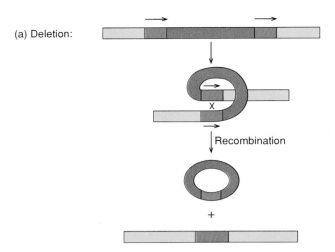
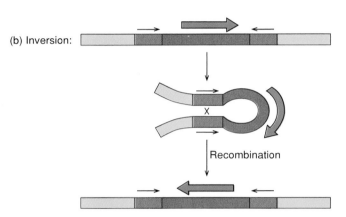

Deletion and Inversion Promoted by Transposons
Figure 12.6

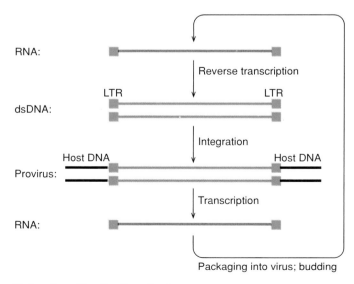

Retrovirus Replication Cycle
Figure 12.10

(a) F-plasmid transfer:

(b) F⁺ → Hfr conversion:

step 1   step 2

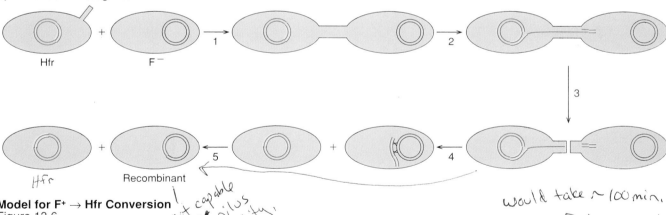

(c) Hfr transfer of host genes:

Model for F⁺ → Hfr Conversion
Figure 13.6

not capable of F- pilus, or F+ fertility, b/c doesn't have all of chromosome replicated

would take ~100 min. for all of chromosome to pass over bridge. b/c of tension, usually breaks before this.

these are steps 1 & 2 from part b on Fig. 13.6. It is showing what different ways it might join

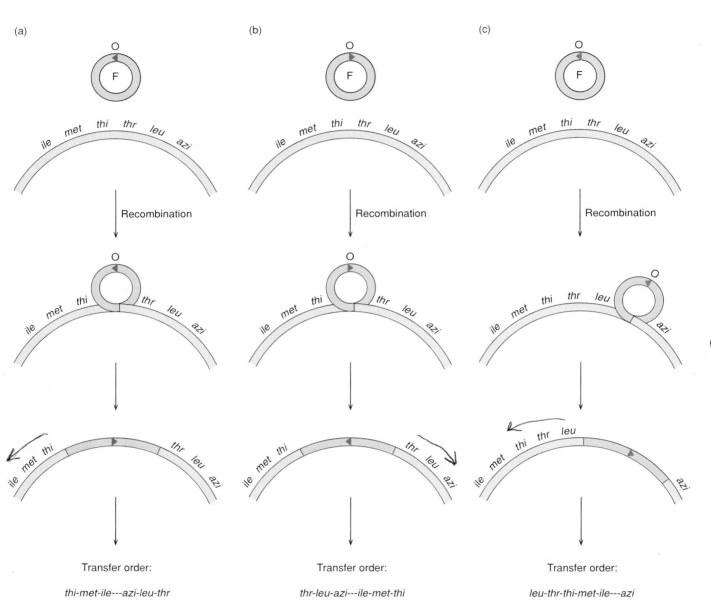

The F Plasmid Portion of an Hfr Chromosome Determines the Direction and Order of Host Gene Transfer
Figure 13.8

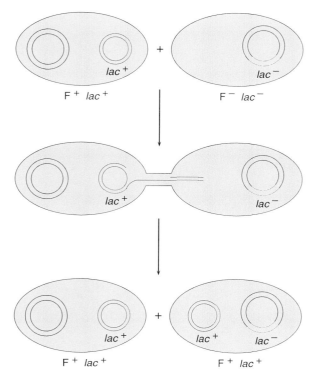

Transfer of F-*lac* to an F⁻ *lac*⁻ Cell
Figure 13.12

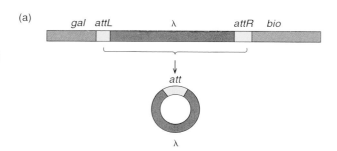

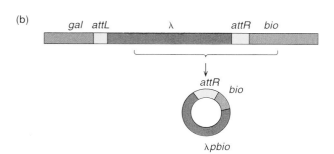

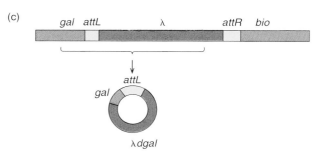

Creation of λ Transducing Phages
Figure 13.15

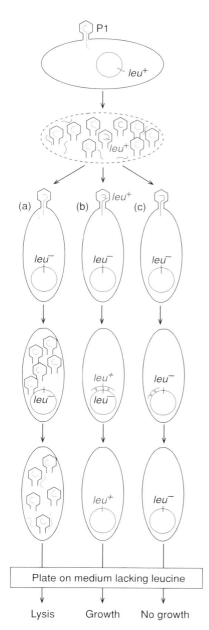

Generalized Transduction
Figure 13.16

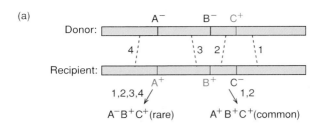
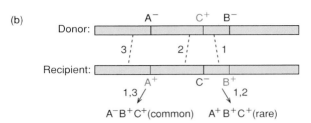

Three-Factor Cross to Determine the Order of Genes A, B, and C
Figure 13.18

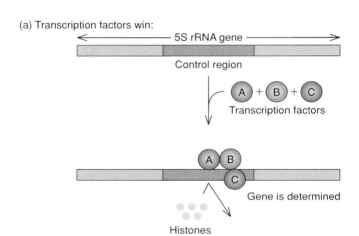

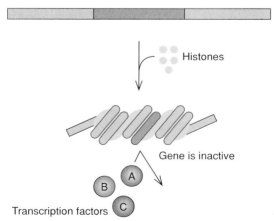

The Race between Transcription Factors and Histones for the 5S rRNA Control Region
Figure 14.11

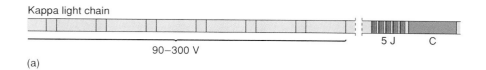

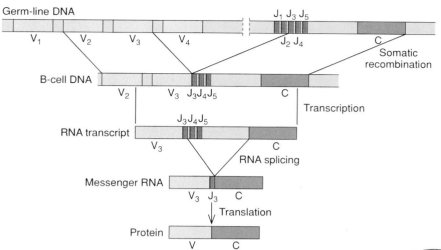

Rearrangement of an Antibody Light Chain Gene
Figure 14.23

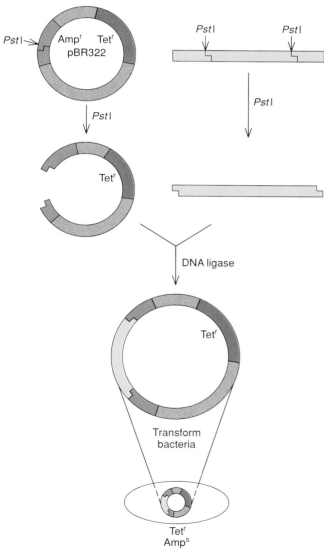

Cloning Foreign DNA Using the *Pst*I Site of pBR322
Figure 15.3

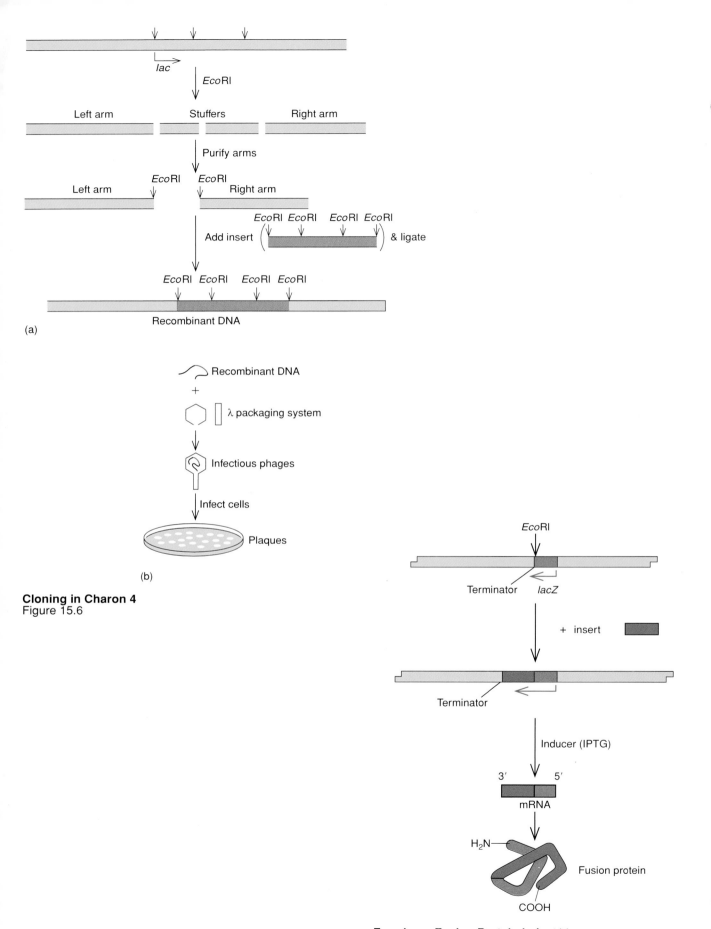

Cloning in Charon 4
Figure 15.6

Forming a Fusion Protein in λ gt11
Figure 15.15

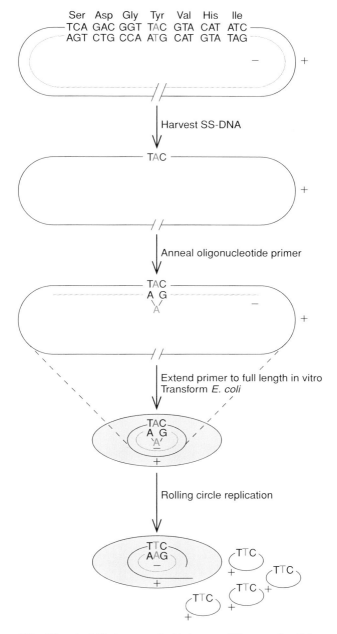

Site-Directed Mutagenesis Using an Oligonucleotide
Figure 15.17

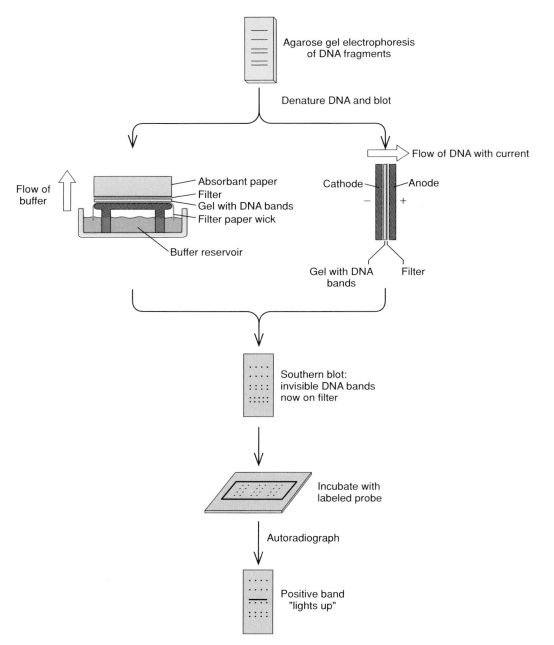

Southern Blotting
Figure 15.18

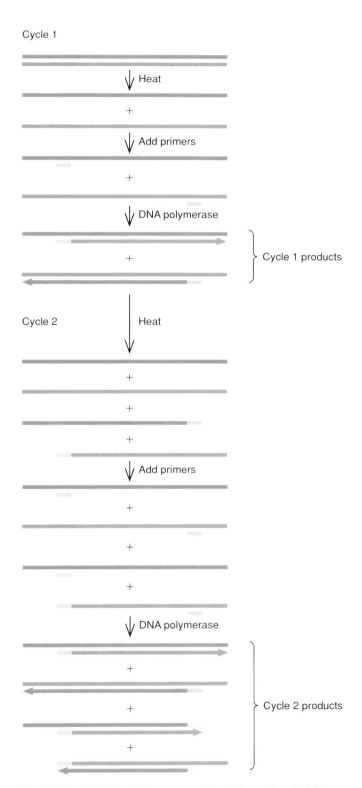

Amplifying DNA by Polymerase Chain Reaction (PCR)
Figure B15.1.2.

(a) Primer extension reaction:

```
—//——————————————TACTATGCCAGA———————————//—
   ————————————————
      21-base primer
```

Replication with ddTTP

```
—//——————————————TACTATGCCAGA———————————//—
   ——————————————————ATGAT
          (26 bases)
```

(b) Products of the four reactions:

Products of ddA rxn

Template: _____ TACTATGCCAGA ____
(22) _____ A
(25) _____ ATGA
(27) _____ ATGATA

Products of ddC rxn

Template: _____ TACTATGCCAGA ____
(28) _____ ATGATAC
(32) _____ ATGATACGGTC

Products of ddG rxn

Template: _____ TACTATGCCAGA ____
(24) _____ ATG
(29) _____ ATGATACG
(30) _____ ATGATACGG

Products of ddT rxn

Template: _____ TACTATGCCAGA ____
(23) _____ AT
(26) _____ ATGAT
(31) _____ ATGATACGGT
(33) _____ ATGATACGGTCT

(c) Electrophoresis of the products:

```
    ddA  ddC  ddG  ddT
                      ┤T
                      ┤C
                      ┤T
                      ┤G
                      ┤G
                      ┤C
                      ┤A
                      ┤T
                      ┤A
                      ┤G
                      ┤T
                      ┤A
```

The Sanger Dideoxy Method of DNA Sequencing
Figure 15.22

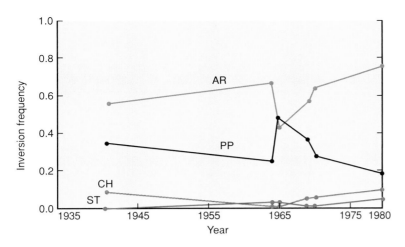

Frequencies of Inversions on the Third Chromosome in *Drosophila pseudoobscura* That Have Taken Place in the Capitan Area of New Mexico
Figure 16.2

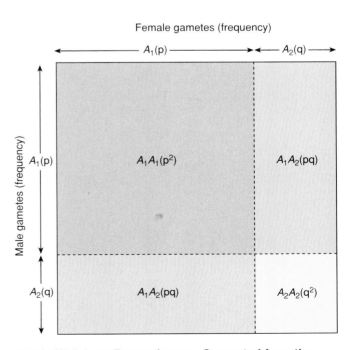

Hardy-Weinberg Proportions as Generated from the Random Union of Gametes, Using a Unit Square
Figure 16.6

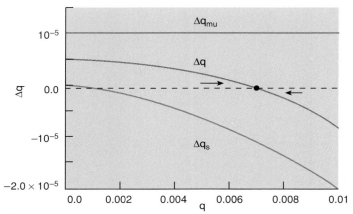

The Change in Allelic Frequencies Where Both Mutation to and Selection Against Recessives Occur
Figure 17.1

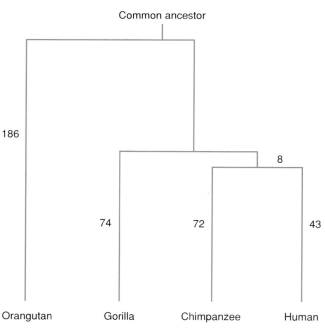

A Phylogeny of Humans, Chimpanzees, Gorillas, and Orangutans, Based on Their Base Sequences in the β-Globin Region
Figure 17.10

CREDITS

Photograph

Fig. 7.12 From John Cairns *Symposia on Quantitative Biology* (Cold Spring Harbor) 28:44, 1963.

Line Art

Fig. 2.7 From: *An Introduction to Genetic Analysis* 3/e by Suzuki, Griffiths, Miller and Lewontin. Copyright © 1986 by W. H. Freeman and Company. Used with permission.

Fig. 4.40 From J. J. Yunis and O. Prakash, *Science,* Vol. 215, March 19, 1982. Copyright © 1982 by the American Association for the Advancement of Science, Washington, DC. Reprinted by permission of the publisher and authors.

Fig. 5.13 From: *An Introduction to Genetic Analysis* 3/e by Suzuki, Griffiths, Miller and Lewontin. Copyright © 1986 by W. H. Freeman and Company. Used with permission.

Fig. 6.14 B Reprinted from *Nature,* vol. 171, p. 737. Copyright © 1953 Macmillan Magazines, Ltd. Reprinted by permission of the publisher and authors.

Fig. 6.14 C Reprinted from *Nature,* vol. 175, p. 834. Copyright © 1955 Macmillan Magazines, Ltd. Reprinted by permission of the publisher and authors.

Fig. 9.6 From Widom and Klug, in *Cell,* 43:210, 1985. Copyright © 1985 Cell Press, Cambridge, MA. Reprinted by permission of the publisher and authors.

Fig. 10.11 Reprinted from *Nature,* vol. 331, p. 225. Copyright © 1988 Macmillan Magazines, Ltd. Reprinted by permission of the publisher and author.

Fig. 10.18 From Gary Quigley and Alexander Rich, *Science,* 185:436, 1974. Copyright © 1974 by the American Association for the Advancement of Science, Washington, DC. Reprinted by permission of the publisher and authors.